THE AGE OF EXPERT TESTIMONY

S0-BYL-394

SCIENCE IN THE COURTROOM

Report of a Workshop

Science, Technology, and Law Panel

Policy and Global Affairs

NATIONAL RESEARCH COUNCIL
Washington, D.C.

NATIONAL ACADEMY PRESS   2101 Constitution Avenue, N.W.   Washington, D.C. 20418

This report was supported by a gift from Procter & Gamble, the Federal Judicial Center, and The National Academies Endowment Fund. Any opinions, findings, conclusions, or recommendations expressed in this publication are those of the author(s) and do not necessarily reflect the views of the organizations or agencies that provided support for the project.

International Standard Book Number 0-309-08310-9

Additional copies of this report are available from National Academy Press, 2101 Constitution Avenue, N.W., Lockbox 285, Washington, D.C. 20055; (800) 624-6242 or (202) 334-3313 (in the Washington metropolitan area); Internet, http://www.nap.edu

Printed in the United States of America

THE NATIONAL ACADEMIES

National Academy of Sciences
National Academy of Engineering
Institute of Medicine
National Research Council

The **National Academy of Sciences** is a private, nonprofit, self-perpetuating society of distinguished scholars engaged in scientific and engineering research, dedicated to the furtherance of science and technology and to their use for the general welfare. Upon the authority of the charter granted to it by the Congress in 1863, the Academy has a mandate that requires it to advise the federal government on scientific and technical matters. Dr. Bruce M. Alberts is president of the National Academy of Sciences.

The **National Academy of Engineering** was established in 1964, under the charter of the National Academy of Sciences, as a parallel organization of outstanding engineers. It is autonomous in its administration and in the selection of its members, sharing with the National Academy of Sciences the responsibility for advising the federal government. The National Academy of Engineering also sponsors engineering programs aimed at meeting national needs, encourages education and research, and recognizes the superior achievements of engineers. Dr. Wm. A. Wulf is president of the National Academy of Engineering.

The **Institute of Medicine** was established in 1970 by the National Academy of Sciences to secure the services of eminent members of appropriate professions in the examination of policy matters pertaining to the health of the public. The Institute acts under the responsibility given to the National Academy of Sciences by its congressional charter to be an adviser to the federal government and, upon its own initiative, to identify issues of medical care, research, and education. Dr. Kenneth I. Shine is president of the Institute of Medicine.

The **National Research Council** was organized by the National Academy of Sciences in 1916 to associate the broad community of science and technology with the Academy's purposes of furthering knowledge and advising the federal government. Functioning in accordance with general policies determined by the Academy, the Council has become the principal operating agency of both the National Academy of Sciences and the National Academy of Engineering in providing services to the government, the public, and the scientific and engineering communities. The Council is administered jointly by both Academies and the Institute of Medicine. Dr. Bruce M. Alberts and Dr. Wm. A. Wulf are chairman and vice chairman, respectively, of the National Research Council.

SCIENCE, TECHNOLOGY, AND LAW PANEL

ERIC S. LANDER, (NAS/IOM), Member, Whitehead Institute for Biomedical Research, Professor of Biology, MIT, Director, Whitehead Institute/MIT Center for Genome Research, and Geneticist, Massachusetts General Hospital, Massachusetts Institute of Technology, Cambridge, Mass.

PATRICK A. MALONE, Partner, Stein, Mitchell & Mezines, Washington, D.C.

RICHARD A. MESERVE, Chairman, Nuclear Regulatory Commission, Washington, D.C.

ALAN B. MORRISON, Director, Public Citizen Litigation Group, Washington, D.C.

HARRY J. PEARCE, Chairman, Hughes Electronics Corporation, El Sagamundo, Calif.

HENRY PETROSKI, (NAE), A.S. Vesic Professor of Civil Engineering and Professor of History, Duke University, Durham, N.C.

CHANNING R. ROBERTSON, Ruth G. and William K. Bowes Professor, School of Engineering, and Professor, Department of Chemical Engineering, Stanford University, Palo Alto, Calif.

PAMELA ANN RYMER, Circuit Judge, U.S. Court of Appeals for the Ninth Circuit, Pasadena, Calif.

STAFF OF THE SCIENCE, TECHNOLOGY, AND LAW PROGRAM

ANNE-MARIE MAZZA, Director
SUSIE BACHTEL, Staff Associate
MAARIKA LIIVAK, Christine Mizrayan Intern
KEVIN WHITTAKER, Christine Mizrayan Intern
KIRSTEN MOFFATT, Consultant
ALAN ANDERSON, Consultant Writer

Preface

A dozen years ago, the Carnegie Corporation of New York established the Commission on Science, Technology, and Government to explore the increasing importance of scientific understanding to decision making in government. Reports issued under the auspices of that commission called attention to the interdependence of science and all three branches of the federal government, including its legal rules and institutions.

Events over the past decade have confirmed the importance of the Carnegie Commission's work by dramatizing the connections, and sometimes the tensions, between science and the law. In response to these connections and tensions, the National Academies created the Science, Technology and the Law Program (STL) in 1998 to more fully explore these issues. The STL Program initiated its work with a two-day meeting in March 2000.

A major topic of that meeting, Scientific Evidence, was explored more fully in a workshop on September 7, 2000. This topic has drawn substantial current attention because of three recent Supreme Court cases that address the admissibility of scientific and other technical evidence in civil litigation.

The purpose of the Scientific Evidence Workshop was to air fully all points of view about the controversial issue of admitting scientists and their testimony into the courtroom. The goal of the workshop was not to reach conclusions or recommendations, but to hear and consider all viewpoints from both legal and scientific leaders with long experience in the field of expert testimony. The current report attempts to be faithful to that original intention by summarizing the proceedings and organizing them by topic; it does not advocate the kinds of specific recommendations commonly seen in reports of the National Academies. It should be pointed out that, while workshop participants did hold strong opinions about the admissibility of evidence and expert testimony, this summary does not attribute views to specific speakers.

Issues of admissibility often become extremely complex in tort litigation, notably in the "toxic tort" litigation where a plaintiff may claim damage by a chemical or product produced by the defendant. Thus, it is anticipated that defendants will generally seek restrictions on the evidence that is admitted to establish liability, while plaintiffs will prefer more lenient standards for admitting such evidence. These points of view, which are central to issues of admissibility, can be perceived throughout the workshop summary.

In light of the difficulties of deciding how to admit expert testimony in cases that often have great economic and personal consequences, the legal system itself must work as effectively as possible. Thus a primary goal of this workshop was to seek ways to improve communication between scientists, lawyers, judges, and juries. It is important to inform the scientific community about how and why science is presented in a certain manner in court, revealing why some points of view that find their way into the courtroom may seem "extreme" or outside the prevailing scientific community consensus. It is equally important to inform the legal community about the scientific method, the scientific concept of the "truth," and the reasons why some of the most qualified scientists may be reluctant to testify in a courtroom. To these ends, this report does discuss a series of techniques—such as court-appointed witnesses, technical advisors, short courses, and other educational tools—that are being used to improve communication and understanding between the worlds of scientists and lawyers. Improved communication and understanding are fundamental to more effective litigation in an age when science and technology underlie many of the customs, decisions, and assumptions of our culture.

Acknowledgments

The Science, Technology, and Law Panel wishes to acknowledge the many fine contributions of the speakers attending the workshop. We especially wish to thank
Margaret A. Berger, Suzanne J. and Norman Miles Professor of Law, Brooklyn Law School; Donald N. Bersoff, Professor of Law and Director of Law and Psychology Program, Villanova School of Law; M. Gregg Bloche, Professor of Law, Georgetown University Law School and Adjunct Professor of Public Health, The Johns Hopkins University; Paul D. Carrington, Harry R. Chadwick Senior Professor of Law, Duke University Law School; Joel E. Cohen, Abby Rockefeller Mauze Professor, and Head, Laboratory of Populations, The Rockefeller University and Columbia University; The Honorable Andre M. Davis, Judge, U.S. District Court for the District of Maryland; Shari Diamond, Professor of Law and Psychology, Northwestern University Law School; David A. Freedman, Professor of Statistics, University of California, Berkeley; Bernard D. Goldstein, Director, Environmental and Occupational Health Sciences Institute, UMDNJ–Robert Wood Johnson Medical School; Leon Gordis, Professor of Epidemiology, School of Public Health, Professor of Pediatrics, School of Medicine, The Johns Hopkins University; Michael H. Gottesman, Professor of Law, Georgetown University Law Center; Michael H. Hoeflich, Dean and John H. and John M. Kane Professor of Law, University of Kansas School of Law; Sheila Jasanoff, Professor of Science and Public Policy, John F. Kennedy School of Government, Harvard University; Gina Kolata, Science Desk, *The New York Times*; Richard A. Levie Principal,

ADR Associates, L.L.C. and Senior Judge, Superior Court of the District of Columbia (ret.); Patrick A. Malone, Stein, Mitchell & Mezines; David Ozonoff, Professor and Chair, Department of Environmental Health, School of Public Health, Boston University; Marsha Rabiteau, Counsel, The Dow Chemical Company; Channing R. Robertson, Ruth G. and William K. Bowes Professor, Department of Chemical Engineering, Stanford University; Jon Samet, Professor and Chair, Department of Epidemiology, School of Public Health, The Johns Hopkins University; and William B. Schultz, Deputy Assistant Attorney General, Civil Division, U.S. Department of Justice.

This report has been reviewed in draft form by individuals chosen for their diverse perspectives and technical expertise in accordance with procedures approved by the NRC's Report Review Committee. The purpose of this independent review is to provide candid and critical comments that will assist the institution in making its published report as sound as possible and to ensure that the report meets institutional standards for objectivity and evidence. The review comments and draft manuscript remain confidential to protect the integrity of the deliberative process.

We wish to thank the following individuals for their review of this report: Arthur Bryant, Trail Layers for Public Justice, Oakland, California; Gilbert S. Omenn, University of Michigan Health System, Ann Arbor; Ernie Rosenberg, Soap and Detergents Association, Washington, D.C.; and Barbara H. Rothstein, U.S. District Court of the Western District of Washington, Seattle.

Although the reviewers listed above have provided many constructive comments and suggestions, they were not asked to endorse the report nor did they see the final draft of the report before its release. The review of this report was overseen by Morris Tanenbaum, who was responsible for making certain that an independent examination of this report was carried out in accordance with institutional procedures and that all review comments were carefully considered. Responsibility for the final content of this report rests entirely with the authoring committee and the institution.

In addition, the Panel also wishes to acknowledge the work of the Science, Technology, and Law Program staff: Anne-Marie Mazza, Susie Bachtel, Maarika Liivack, Kevin Whittaker, Kirsten Moffatt, and consultant, Alan Anderson.

Donald Kennedy and Richard A. Merrill
Co-chairs

Contents

1

Introduction

"In this age of science, we must build legal foundations that are sound in science as well as in law. Scientists have offered their help. We in the legal community should accept that offer. We are in the process of doing so."
—Associate Justice Stephen Breyer
"Introduction" in *Reference Manual on Scientific Evidence,*
Second Edition (Federal Judicial Center, 2000)

As society has come increasingly to depend on science and technology, more scientists, engineers, and physicians are asked to testify in courts of law on technical questions in their fields of expertise. They may be asked to provide facts in some cases and opinions in others, but their presence as "expert witnesses" has become familiar in cases dealing with disputed issues in tissue testing (do the blood samples match?), toxicology (does a certain chemical cause chromosomal abnormalities?), epidemiology (is another chemical associated with increased incidence of lung cancer?), engineering (why did a certain device fail?), and other subjects of technical complexity.

Scientists and other technical experts may be asked to testify in several kinds of civil litigation. These include product liability litigation, toxic tort cases, medical malpractice suits, and challenges to regulations that question the adequacy of the scientific underpinning of an administrative case decision or regulation issued by a regulatory agency. There are also occasional opportunities for court review of agency decisions in some cases. Such decisions are generally subject to three types of interrelated tests: (1) the statutory wording that governs the level of scientific or engineering certainty about toxicity or product risk; (2) the specific level of protectiveness required (e.g., "ample margin of safety" or the absence of "unreasonable risk"); and (3) whether the findings are governed by a legal test of deference to an agency decision ("preponderance of the evidence" vs. "arbitrary and capricious").[1]

The courts apply no single level of rigor for the science or other expertise used as evidence in civil litigation. The rigor applied by the court may vary as a function of the jurisdiction of the court (federal, state, administrative law judge whether the case is governed by common law or statute) and of the stakes involved. Higher stakes (e.g., major injuries, massive recalls) sometimes appear to require more evidence and a higher level of proof of causation. In addition, juries may be more skeptical of evidence if a verdict will result in an enormous award or threaten the viability of a major institution. The rigor of scientific evidence may vary with the purpose of the inquiry; a lower level of rigor may be required to simply list a substance as "toxic" for reporting purposes (e.g., for the Toxic Release Inventory), a higher level to impose regulatory restrictions, an even higher level to award huge damages, and a higher level still to ban a substance from use.[2]

One arena that has proven especially troublesome for courts is the question of causation in so-called toxic tort cases, which might be considered as a subspecies of product liability litigation.[3] Because many of these cases turn on points of technical complexity, expert witnesses are often required to provide testimony. Much is at stake—the viability of firms or whole industries, the ability for injured parties to receive adequate compensation, and costly litigation burdens and settlement sums. Therefore, courts have tried hard to improve their standards for admitting and weighing scientific and technical evidence.

---

[1]See, e.g., *Chevron U.S.A., Inc. v. Natural Resources Defense Council, Inc.*, 467 U.S. 837 (1984), *Christensen v. Harris County*, 529 U.S. 576 (2000) . See also Ernest Gellhorn & Paul Verkuil, Controlling Chevron-Based Delegations, 20 Cardozo L. Rev. 989 (1999), *U.S. v. Mead Corp.*, 533 U.S. (2001).

[2]There are also important distinctions among rigor (the reliability of evidence), accuracy (test variability), and precision (the level of accuracy to which a test can measure).

[3]The prevalence of toxic tort litigation, which seeks to resolve claims by individuals that they have been injured by exposure to chemicals or other products, has grown rapidly in recent years, amid considerable controversy. Some people criticize toxic tort litigants for seeking to impose liability for exposures that carry little or no risk to individuals (e.g., air pollution), especially when compared to other, larger risks under a person's voluntary control (e.g., driving, smoking). The law reflects this distinction, however, and finds relatively low risks that are involuntary to be actionable or sufficient to warrant regulation.

The resolution of toxic tort cases may affect large classes of people or ways of performing critical functions. As Justice Breyer has pointed out, "[M]odern life, including good health as well as economic well-being, depends upon the use of artificial or manufactured substances, such as chemicals. And it may, therefore, prove particularly important to see that judges . . . help assure that the powerful engine of tort liability, which can generate strong financial incentives to reduce, or to eliminate, production, points towards the right substances and does not destroy the wrong ones. It is, thus, essential in this science-related area that the courts administer the Federal Rules of Evidence in order to achieve the 'end[s]' that the Rules themselves set forth, not only so that proceedings may be 'justly determined,' but also so 'that the truth may be ascertained.' " [4]

At the same time, many scientific, engineering, and medical experts who have been asked to testify in the courtroom have experienced tension and occasionally frustration. Some feel that they are unable to communicate their knowledge in sufficient detail to nonscientists; others find their evidence, their expertise, and even their careers under attack.

Another cause of experts' frustration transcends the high stakes of litigation. The role of scientists in the courtroom turns out, as one would expect, to differ in significant respects from their role in the laboratory. For example, in the courtroom, scientists are often asked for a yes-or-no answer to a question in which they discern shades of gray. In addition, inherent differences between the cultures of law and science are magnified under courtroom conditions, especially in the context of cross-examination. There is anecdotal evidence that these differences may dissuade scientists and engineers from stepping forward in cases where the nation needs the benefit of their expertise.

BRINGING LAWYERS AND SCIENTISTS TOGETHER

For these and other reasons, the National Academies, in their role representing the scientific, engineering, and medical communities, have taken a strong interest in the preparation, presentation, and handling of expert testimony. That interest crystallized several years ago when the case *Daubert v. Merrell Dow Pharmaceuticals, Inc.* reached the Supreme Court.[5] The National Academies were moved to join with the American Association for the Advancement of Science (AAAS) to file an *amici curiae* brief in a case that has done much to update the role of scientific expert

---

[4]*General Electric Co. v. Joiner*, 118 S.Ct. 512, 520 (1997) (J. Breyer, concurring), citing Fed. Rule. Evid. 102.

[5]*Daubert v. Merrell Dow Pharmaceuticals, Inc.*, 113 S. Ct. 2786 (1993).

witnesses and the admissibility of scientific evidence. The Court largely adopted the argument of the brief that scientific evidence should be evaluated by the standards of the scientific community. Justice Blackmun noted that "there are no certainties in science" and quoted the AAAS/NAS *amici* brief : "Science is not an encyclopedic body of knowledge about the universe. Instead, it represents a *process* for proposing and refining theoretical explanations about the world that are subject to further testing and refinement."[6]

One of the outcomes of *Daubert* (pronounced Dow-bert) is that judges have considerable responsibility in understanding and acting on questions of scientific complexity that may have their roots in such fields as epidemiology, toxicology, and molecular genetics. To fulfill this responsibility, judges—at least federal trial judges—are asked to serve a "gatekeeping" role in deciding whether the expert testimony is sufficiently reliable to be presented at trial. In doing so, judges, few of whom have technical training, are asked to exercise a degree of expertise themselves in grappling with cause-and-effect issues on which scientific experts themselves may disagree. Of equal concern are the differences in language and culture between science and law that are heightened in the courtroom.

In order to survey the difficulties produced by this considerable responsibility, the Science, Technology, and Law Program of the National Academies invited individuals from the legal, scientific, and engineering communities, most of whom had considerable experience with expert testimony, to a day-long workshop to seek a full range of opinions. The workshop was not intended to seek a consensus. Instead it sought to air differences and explore emerging problems. In order to achieve this purpose, the National Academies invited experts representing a broad spectrum of opinion as to what level of scientific validation should be required to be admitted as evidence in civil cases.

The following summary attempts to capture the main points of the day's discussion. While such a brief summary necessarily omits much, its purpose is to illuminate the main features of the landscape in which scientists and lawyers find themselves in light of *Daubert* and the challenge of identifying objective and unbiased scientific experts for court-appointed roles. Among those features are the nature of expert evidence, the rules of evidence as applied to science, the scientific method and its application in law, and the many difficulties in reaching conclusions about causation when experts disagree.

---

[6]Brief for the American Association for the Advancement of Science and the National Academy of Sciences as *Amici-Curiae* 7-8.

2

The Supreme Court Trilogy

"I believe strongly that the Supreme Court got it [*Daubert*] right. . . . I think the legal system will get it right."

—Judge Andre Davis

Beginning in 1993, a series of Supreme Court decisions altered the evidentiary landscape for scientific issues and experts. Prior to 1993, the common standard for admitting expert evidence in federal courts was set by *Frye v. United States*, a federal appellate court ruling of 1923. This ruling held that expert opinion based on a scientific technique is only admissible if the technique is "generally accepted" in the relevant scientific community.[7] General acceptance, however, was sometimes determined on the basis of the testimony of a self-validating expert. The "Supreme Court trilogy," discussed below, encourages trial judges to decide admissibility not solely on this standard of consensus or general acceptance, but on whether the testimony is grounded in the principles and methods of a particular field.

DAUBERT V. MERRELL DOW PHARMACEUTICALS (1993)

Daubert began when the parents of two minor children alleged that the children's birth defects had been caused by the mothers' prenatal

---

[7]*Frye v. United States*, 293 F. 1013, 1014 (D.C. Cir. 1923).

ingestion of Bendectin, a prescription anti-nausea drug marketed by Merrell Dow. A district court accepted the affidavit of an expert who reviewed the scientific literature on the subject and concluded that maternal use of Bendectin had not been shown to be a risk factor for human birth defects.

The court disallowed the evidence of eight other experts who argued that Bendectin could indeed cause birth defects. The evidence of these experts was based on animal studies, chemical structure analyses, and the unpublished "reanalysis" of previously published human statistical studies. The court, citing *Frye*, stated that expert opinion that is not based on epidemiological evidence is not admissible to establish causation. Therefore, the evidence of the eight experts, ruled the court, was not admissible because it was not "sufficiently established to have general acceptance in the field to which it belongs." The Court of Appeals for the Ninth Circuit agreed.

The Supreme Court vacated and remanded this judgment for further proceedings, ruling that the enactment of the Federal Rules of Evidence in 1975 superceded the general acceptance test of the then-50-year-old *Frye* test. The Court pointed to Rule 702, which permits experts to testify on matters of "scientific, technical or other specialized knowledge" if it will assist the judge or jury to understand the evidence. The rule also places "appropriate limits" (in the words of Justice Blackmun) on the admissibility of purportedly scientific evidence by assigning "to the trial judge the task of ensuring that an expert's testimony both rests on a reliable foundation and is relevant to the task at hand. Pertinent evidence based on scientifically valid principles will satisfy those demands."[8]

The Court reasoned that, if expert testimony is offered in the form of "scientific . . . knowledge," that testimony must be based on a reliable scientific methodology. Under the Rules, the trial judge is now responsible for assessing both its relevance and reliability. The subject of an expert's testimony must, according to *Daubert*, be "scientific . . . knowledge," and this knowledge must be grounded in the methodology and reasoning of science.

This decision imposes a considerable "gatekeeping" responsibility on federal trial judges and provides guidance, in the form of four suggested factors, for screening expert scientific testimony. The four *"Daubert* criteria" for evaluating the admissibility of expert testimony are:

1. whether the theories or techniques upon which the testimony relies are based on a testable hypothesis;

---

[8]*Daubert v. Merrell Dow.*

2. whether the theory or technique has been subjected to peer review;
3. whether there is a known or potential rate of error associated with the method; and
4. whether the method is generally accepted in the relevant scientific community.

These criteria are neither mandatory nor exclusive, as elaborated in *Joiner* and *Kumho* below.

GENERAL ELECTRIC CO. V. JOINER (1997)

In the second case of the trilogy, decided four years later, the Court reinforced the responsibility of the trial judge to screen scientific evidence for the standards of reliability and relevance.

Robert Joiner, an electrician in the Water & Light Department in Thomasville, Ga., filed suit in state court, alleging that his small-cell lung cancer had been caused by on-the-job exposure, beginning in 1973, to polychlorinated biphenyls (PCBs) contained in the coolant of electrical transformers. The transformers and the dielectric fluid used as a coolant were manufactured by General Electric and Westinghouse Electric.

The Court repeated its opinion that the "austere" *Frye* standard of "general acceptance" had not been carried over into the Federal Rules of Evidence. In the *Joiner* case, however, the Court upheld the trial court's exclusion of certain evidence—not because it was "unacceptable," but because it was not relevant to the particular case. In particular, it found that the animal studies offered as evidence were too ". . . dissimilar to the facts presented in this litigation," and that the four epidemiological studies "were not a sufficient basis for the experts' opinions" regarding this plaintiff's illness.

For example, said the court, ". . . the animal studies involved infant mice that had developed cancer after being exposed to PCBs. The infant mice in the studies had had massive doses of PCBs injected directly into their peritoneums or stomachs. Joiner was an adult human being whose alleged exposure to PCBs was less than the exposure in the animal studies." In addition, ". . . the cancer that these mice developed was alveologenic adenomas; Joiner had developed small-cell carcinomas." Finally, "No study demonstrated that adult mice developed cancer after being exposed to PCBs. Weaknesses were also found in each of the epidemiological studies.

The Court made clear that in weighing admissibility the judge has the responsibility to ensure that testimony meets the criteria of relevance and reliability. The Court determined that the studies on which the expert's opinions were based were not sufficiently tied to the issues in the litigation, noting that "[N]othing in either *Daubert* or the Federal Rules of Evidence requires a district court to admit opinion evidence which is connected to existing data only by the *ipse dixit* [personal opinion] of the expert. A court may conclude that there is simply too great an analytical gap between the data and the opinion proffered." The Court ruled that the district court judge had not abused her discretion in excluding the scientific evidence.

KUMHO TIRE CO. LTD. V. CARMICHAEL (1999)

In its ruling in *Kumho*, two years after *Joiner*, the Court extended its reasoning beyond the "scientific" evidence of *Daubert* to all expert testimony based on "skill- or experience-based observation." The four *Daubert* factors may be relevant to such evidence, but are not essential; other factors may be appropriate, given the nature of the testimony and the particular circumstances of the case. Thus does *Kumho* complete the *"Daubert* trilogy."

The evidence at issue in the case of *Kumho* was the testimony of an expert in tire failure analysis. The suit began when a worn rear tire of a minivan driven by Patrick Carmichael blew out. In the accident that followed, one of the passengers died, and others were severely injured. The Carmichaels brought suit against the manufacturer, Kumho Tire, claiming that the tire was not merely worn but defective, resting their case largely on the testimony of a tire expert. The District Court, referring to the *Daubert* criteria, disallowed the testimony of the tire expert because it was not considered sufficiently reliable or "scientific."

The Supreme Court agreed that the testimony should be excluded, emphasizing that technical evidence need not be judged by standards appropriate to research results that are published in the scientific literature: "We conclude that *Daubert*'s general holding—setting forth the trial judge's general 'gatekeeping' obligation—applies not only to testimony based on 'scientific' knowledge, but also to testimony based on 'technical' and 'other specialized' knowledge."[9] The Court made it clear that evidence is to be measured against the level of intellectual rigor that characterizes the practice of an expert in the relevant field outside the courtroom, thereby expanding on the four criteria suggested under *Daubert* for "scientific" evidence.

A significant outcome of the opinion was to expand the trial judge's gatekeeping role to all expert testimony, even testimony that might not be considered "scientific" in a strict sense. In this case, the Court acknowl-

---

[9]Justice Breyer is quoting from Rule 702 of the Federal Rules of Evidence.

edged that the tire expert was appropriately qualified because of his long experience and expertise, but upheld exclusion of his testimony. The tire expert's testimony, wrote Justice Breyer, was properly excluded because the expert failed to employ standards used by similar experts who make such assessments outside the courtroom.

Following the *Kuhmo* decision, the Federal Court codified this emerging body of law by amending Rule 702 of the Federal Rules of Evidence, as follows: If scientific, technical, or other specialized knowledge will assist the trier of fact to understand the evidence or to determine a fact in issue, a witness qualified as an expert by knowledge, skill, experience, training, or education, may testify thereto in the form of an opinion or otherwise if (1) the testimony is based upon sufficient facts or data, (2) the testimony is the product of reliable principles and methods, and (3) the witness has applied the principles and methods reliably to the facts in case.

3

The Nature of Expert Evidence

"My favorite novelist is Trollope. He was once a fact witness in a court case. The cross-examination was something like the following: 'Now Mr. Trollope, you're the author of 38 books, aren't you?' 'Yes.' 'And there isn't a word of truth in any of them, is there? ' "

— David Freedman

As the number of legal cases raising complex scientific and technical issues has increased, along with the financial consequences to individuals and businesses, courts have struggled to find a clear standard by which to admit expert evidence. Some critics have attempted to begin by posing the existence of a line between science and "junk" science. Claims that junk science has infected the nation's courtrooms are common, even though few people agree on a definition for the term. One participant suggested that "junk" is a general term used to describe evidence that favors plaintiffs in product liability or toxic tort litigation. Another suggested that the term was sometimes applied to scientists employed or funded by corporations and who testify on behalf of those corporations. More broadly, however, the existence of the term reflects some confusion and even cynicism on the part of both expert witnesses and lawyers about the value and objectivity of evidence.

Making expert evidence comprehensible to laypeople serving on juries constitutes a significant challenge to the court, argued one participant. Current trial procedures, he said, have arisen from a need to resolve conventional disputes of all kinds, most of which have little or no "technical" content. In general, a trial is an educational and inferential process in which evidence is heard, weighed against other evidence, and applied to the circumstances of the case to determine a verdict.

Admitting scientific and other expert evidence that is unfamiliar to the average juror presents a fundamental shift in this paradigm, stated one participant. Further, noted the participant, there is a sharp difference between "conventional" evidence, to which non-experts on a jury may add personal life experiences that aid them in deliberations, and more specialized scientific evidence that "can only be deferred to." One danger seen by this participant is that judges, knowing that jurors may defer to evidence or opinions they do not understand, may be excessively rigorous in excluding expert evidence in order to "protect" jurors.

THE INCOMPLETE NATURE OF EVIDENCE

As much as courts would like to winnow out unsound evidence, however, the task is not an easy one. As one lawyer emphasized, "all evidence is incomplete, and it's always going to be." To require absolute reliability in evidence is to expect an elusive and "non-existent purity in science" where virtually all scientific data might fail the most rigorous tests of reliability and relevance in some degree. This may be inherently unfair to whoever has the burden of proof. In toxic tort cases, the plaintiff has the burden of proof, and a court may question the fairness of asking a plaintiff to suffer because of scientific uncertainty. The lawyer added that virtually all scientific data might fail the most rigorous tests of "reliability" and "relevance" in some way. From a scientific standpoint, it should be apparent that a legal structure that requires the defendant to prove a negative would also be problematic.

Other participants discussed the understandable desire on the part of courts for certainty and "bright lines" by which to decide on the admissibility of evidence. They said it is unfair to require a higher standard of accuracy from expert witnesses in the courtroom than exists in the world of science. Other participants noted that it is unfair to admit evidence that would not stand up to the scrutiny of the scientific community.

WHAT CAUSES AN ABSENCE OF SCIENTIFIC DATA

Courts may lack sufficient knowledge about a given chemical, disease agent, or other issue. The presence or absence of a valid scientific study may by itself determine a verdict in the courtroom. One scientist described a "decision tree" by which to explore the factors that motivate groups to fund certain studies and not others.

In epidemiology, for example, the court must often answer the question, Has an association been demonstrated between a certain chemical and human injury? There are two possible answers. If the answer is yes, the court might decide to compensate the plaintiff. If the answer is no, the question is, Why not?

Again, there are two possible answers: either a study has been done and it shows no association, or a study has not been done. If a study has not been done, the question again is, Why not? Again there are two possibilities: one, a study may not be feasible because of the current state of technology, the availability of funding, or other reasons; or, two, the study could have been done, but it has not.

If such a study is not feasible, the court may ask whether it is fair to expect a plaintiff to suffer in the absence of scientific knowledge and the court may decide to compensate. On the other hand, one could argue that a defendant should not have to pay compensation if there is no sufficient evidence. The court may ask a defendant why such a study was not conducted. If a study has not been done that could have been, perhaps there is a mechanism to do the study now.

This decision tree was offered as a basis for discussion. It also served to remind the workshop participants of how many factors influence the availability of sound evidence.

CAN THE JUDGE AND JURY BE EXPECTED TO UNDERSTAND SCIENTIFIC EVIDENCE?

The *Daubert* trilogy sets high expectations for the ability of trial judges to weigh the evidence of experts. A judge at the workshop summarized the responsibilities of the trial judge who is faced with admitting or rejecting scientific testimony. Under Rule 702, which governs admissibility, the trial judge must make a preliminary assessment of whether the testimony's underlying reasoning and methodology are scientifically valid and can be applied to the facts at issue. The judge may consider whether the theory or technique in question can be (or has been) tested, whether it has been subjected to peer review and publication, its known or potential error rate, the existence and maintenance of standards controlling its operation, and whether it has attracted widespread acceptance within a relevant scientific community.

A district judge has great discretion in determining admissibility of scientific evidence. The consequences of excluding scientific evidence at the trial court level can be great, noted a judge. If the judge should exclude critical scientific testimony and then enter summary judgment (i.e., deliver a decision without a jury trial) against the plaintiff on the grounds of lack of evidence, then that judgment would be given great deference by a court of appeals.[10]

BETTER EQUIPPING COURTS TO UNDERSTAND SCIENCE

A participant suggested several ways to assist judges and jurors in dealing with expert evidence: (1) Experts could be required to produce detailed, written reports, as described (but sometimes not honored) in the new Rule 26(a) of the Federal Rules of Civil Procedure. These reports could be read and digested more carefully than oral testimony, which may be delivered in ambiguous language; (2) judges could explain clearly to jurors the existence of any biases among experts who were providing testimony;[11] (3) juries and judges could be provided with better tools to evaluate expert testimony. At present, he said, jurors often get a "pitiful standard jury instruction which basically tells them that you can evaluate experts based on how confident they seemed of their views." Instead, he advocated a more explicit discussion of some of the criteria that underlie good science. Juries, thus equipped, could then ". . . wrestle more fully with this very elusive, but very human enterprise that we call science."

Another participant suggested more organized programs to help judges and juries deal with the demands of *Daubert*.[12] Steps that might be helpful include improving the educational tools provided to judges, developing model jury instructions, studying jury understanding of technical

---

[10]Note that in his decision on *Joiner*, Chief Justice Rehnquist wrote that the Court of Appeals erred in its review of the exclusion of Joiner's experts' testimony, and wrote: "In applying an overly 'stringent' review to that ruling, it failed to give the trial court the deference that is the hallmark of abuse of discretion review."

[11]The potential for bias may be obvious in cases where a witness is an employee of the defendant, or paid by the defendant. But even witnesses who do not have a pecuniary stake in the outcome may want to bolster a pet theory, attack big business for personal reasons, or create an incentive for more research in which they may play a part. It is not unreasonable to suspect bias in someone who always testifies for the defense or always testifies for the plaintiff in toxic tort cases, regardless of the financial situation. Under existing law the parties are free to explore these issues both through cross-examination of an expert and by extrinsic proof.

[12]There are several programs providing judges with additional assistance: (1)The Court Appointed Scientific Experts (CASE) demonstration project, launched by the AAAS, helps judges identify qualified scientists and engineers to serve as experts to the court; (2) The Private Adjudication Center, Inc., of Duke University School of Law also maintains a registry of independent scientific and technical experts who are willing to provide advice to courts or serve as court-appointed experts; and (3) The Federal Judicial Center, the National Center for State Courts, and the National Judicial College offer programs to acquaint judges with scientific principles and evidentiary standards used to evaluate the admissibility of technical proof.

evidence, and developing standards for expert witnesses. The last step, noted the participant, is controversial on the grounds that standardization might not be desirable or possible among the varying disciplines of science.

A substantial debate was raised over the question of whether juries are asked to do too much in trials of technical complexity, such as patent cases. A judge argued strongly in favor of keeping juries for all trials. "Juries are the bedrock of the system," he said. Rather than replacing juries, he suggested that the responsibility for making trials comprehensible to jurors rests with the judge. "We must make the work of the jurors more doable," he said.

4

The Scientist's Role in the Courtroom

"Here's the bottom line. When you ladies and gentlemen with your M.D.s and Ph.D.s come into a courtroom, you are going to be forced to use the court's language, you are going to have to follow the court's procedures. Or put a little differently, you're going to be playing in my ballpark and by my rules. It will not be a scientific meeting where one presenter gets up after another with their power points and their overheads and presents their paper. They get a few questions and then you move on to the next person. It's a very, very different situation."
— Judge Richard Levie

Many scientists, engineers, and physicians are understandably reluctant to testify in court, and the workshop spent considerable time discussing this issue. From the point of view of the scientist, the courtroom is "someone else's turf, where the rules are different and unfamiliar." This situation, said one veteran scientific witness, "is a challenge that used to frighten me and continues to worry many of my colleagues as they consider whether to step into the courtroom." Common concerns of scientists are that they will be embarrassed publicly, their results may be misunderstood or used out of context, and that they may be branded as a "hired gun" for one side or another of an issue.

One reason scientists are uncomfortable in the courtroom is that they are neither trained in nor comfortable with the formalism of the legal adversary proceeding as a mechanism to resolve scientific differences. One scientist discussed the modes of debate in science, which traditionally lead to consensus, not victory or defeat. When a group of scientists is asked to address a question, the group eventually recognizes the value of the strongest evidence and opinions. At that point, even if one or a few members of the group are at extreme ends of the bell-shaped curve of opinion, the custom is for all to join in a "consensus truth."

In the courtroom, the goal is not a consensus truth but a definitive decision. Although there may be a consensus in the scientific community about a particular question, this consensus is unlikely to appear in the courtroom. Instead, opposing attorneys search out experts from the tails of the bell-shaped curve so as to strengthen their particular arguments.

Even so, some scientific, engineering, and medical experts feel a responsibility to make themselves available to offer courtroom testimony; some of whom go so far as to define a collective responsibility to do so. Their reasoning is that the best way to provide sound evidence for legal questions is to provide the most qualified experts. "If those of you who are honorable, conscientious, and learned don't show up," said a judge, "then who do you think will show up?"

Another view presented was that both science and the law are human activities and, in certain respects, social constructs. It is not surprising that members of each group are unfamiliar with the culture and "professional myths" of the other. Participating in resolving legal disputes is one way for the two cultures to "untangle those myths" and learn to communicate better.

In discussing ways to encourage the scientific community to improve expert testimony, participants discussed a role for professional societies and for programs of the AAAS and the National Academies. Some societies have, or are considering, codes of practice for this purpose. Other participants doubted that professional societies would have enough courtroom expertise to be helpful and might even, particularly for licensed professions, find difficulty in resolving definitions and issues of practice.

THE PRACTICE OF SCIENCE

In its *Daubert* opinion, the Court repeatedly uses the term "scientific method," a concept discussed at length during the workshop. In general, the participants disliked the idea of a too-exact definition in regard to science. As one participant said, "There is no definition of 'scientificity'."

As suggested previously, the hypotheses and knowledge of science are always evolving, and this includes the science that underlies scientific evidence. An hypothesis can be falsified or disproved, but it cannot be labeled "permanently true," because the knowledge on which it is based is always incomplete. An hypothesis that is tested and found "not to be false" is considered to be corroborated but not proved.

Indeed, the work of science is to test hypotheses. This process may begin with an observation about the world and proceed to the formulation of an hypothesis to explain that observation; this is followed by the performance of experiments and the collection of data to test the hypothesis, and finally by a lengthy process of peer review, publication, and attempts by other researchers to replicate the experiment. Out of this process grows consensus within the scientific community.

As one brief filed in *Daubert* suggested, "Scientific methodology today is based on generating hypotheses and testing them to see if they can be falsified; indeed, this methodology is what distinguishes science from other fields of human inquiry."[13] This statement is followed in the *Daubert* opinion by a quote from Karl Popper, another eminent philosopher of science, who wrote in the 1930s: "[T]he criterion of the scientific status of the theory is its falsifiability or refutability or testability."[14]

Chief Justice Rehnquist was sufficiently perplexed by this assertion to offer a mild dissent: "I defer to no one in my confidence in federal judges, but I'm at a loss to know what is meant when it is said that the scientific status of a theory depends on its falsifiability, and I expect some of them will be confused, too.[15]

Workshop participants found no problem with refutability, or testability. Scientific experiments are published publicly for the purpose of offering others the chance to replicate the results of the experiment and either to refute them or confirm them. They did discuss the notion of falsifiability at some length, however, as a concept that was produced by philosophers of science in the 1930s, 1940s, and 1950s, but which has been incorporated into more sophisticated conceptions of science today.

GOOD SCIENCE, BAD SCIENCE, AND PSEUDO-SCIENCE

Several participants discussed the common public perception that scientists sometimes offer bad or "pseudo" science as evidence in the courtroom. A scientist at the workshop emphasized several reasons for this perception. He said that at one extreme of an imaginary spectrum are the "hard" sciences, such as molecular biology, physics, and chemistry. At the other end of the spectrum are the "pseudo-sciences," such as astrology and numerology. In the middle, he said, are many topics whose

---

[13]Green, "Expert Witnesses and Sufficiency of Evidence in Toxic Substances Litigation: The Legacy of *Agent Orange* and Bendectin Litigation," 86 *Nw. U. L. Rev.* 643 (1992), cited in *Daubert amici curiae* brief of Nicolaas Bloembergen et al.

[14]K. Popper, "Conjectures and Refutations: The Growth of Scientific Knowledge," 37 (5th ed. 1989).

[15]*Daubert*, 61 U.S.L.W. at 4811.

status is less clear, including acupuncture, handwriting analysis, and psychological profiling.

The important point, he said, is that even the pseudo-sciences share features with the harder sciences, but they do not have "enough of the virtues, such as refutability," to elevate them to the status of science. Conversely, even some of the hardest sciences lack some aspects of "scientificity." Quantum mechanics, for example, contains elements that are not testable, intuitive, or conservative. The important point is that there is no crisp distinction or algorithm that divides science, almost-science, and pseudo-science.

"The difficulty of trying to tell science from pseudo-science, or real science from unreal science," said the scientist, "is not a problem to be solved. It's a problem to be gotten over. As the software people say, 'It's not a bug, it's a feature'."

Because scientific areas have different standards for assessing evidence, experts can in good faith disagree over interpretations of data and other evidence. Even though scientists strive for consensus, every respected scientific journal reflects the lively debates that precede consensus. In court, two competent and ethical scientists may disagree in good faith, and such disagreements should not be taken as a sign that bad science is being practiced or that science has failed. Freedom to disagree should be seen as a strength, not a weakness, of science.

Participants also discussed the "huge variation" among the cultures of large scientific communities that may cause confusion in the courtroom. For example, epidemiologists, who deal with probabilities rather than certainties, tend to speak with restraint; pathologists, whose task is to define a particular mechanism of disease or cause, are more willing to draw conclusions.

ASSESSING RISK

One of the hardest jobs scientists are asked to do—in the courtroom or in the laboratory—is to assess the risk to humans of certain chemicals or other agents. The workshop was attended by experts in both epidemiology and toxicology, two fields in which experts are commonly asked to provide evidence of risk.

The task of risk assessment is often complicated by conflicting evidence from experts in these two fields. For example, an epidemiological study of 300,000 petroleum workers showed that the chemical benzene is not associated with non-Hodgkin's lymphoma. However, the toxicological literature, describing workers in China exposed to higher levels of benzene, provides evidence of chromosomal abnormalities that are specific for non-Hodgkins lymphoma. In addition, benzene can cause non-

Hodgkin's lymphoma in laboratory animals. Is benzene, therefore, toxic or benign? Is its toxicity a matter of the exposure level?

Another source of confusion is the limits of epidemiology, in which inferences are applied to populations of persons, not individuals. It is possible to say that cigarette smoking increases the likelihood of lung cancer in large populations, for example, but virtually impossible to say with certainty from epidemiological studies alone whether a particular individual's cancer was caused by smoking.

An epidemiological study is said to be sound when its conclusions meet certain generally accepted criteria. Among these are:

- Consistency: Not just one population, but several populations should show an elevated risk of illness from exposure.
- Strength of association: Stronger associations are more likely to have a causal explanation, because potential biases are less likely to cause the association.
- Temporality: Exposure must precede effect.
- Plausibility: Is a cause plausibly linked with an effect? Is the evidence consistent with the larger body of scientific knowledge that pertains to the topic? Is there a biological mechanism that explains how the changes take place?

Epidemiologists prefer to see their evidence confirmed by all of these criteria, as well as some of the additional Bradford Hill criteria[16] (such as dose response, specificity of the association, and replication). Their application, however, is a matter of judgment, where differences can easily arise.

RELATIVE RISK

One concept that causes confusion in the courtroom is that of relative risk. Here again, the scientist's cautious search for consensus often conflicts with the litigant's need for a prompt decision regarding an individual dispute.

For example, some courts have adopted a policy that a relative risk greater than 2.0 must be shown to establish causality.[17] That is, the court

---

[16]Bradford-Hill, A., "The Environment and Disease: Association or Causation?" Proc. Royal Soc. Med. 58:295 (1966); see also: Bradford-Hill, A., "The Environment and Disease: Association or Causation?" President's Address. Proc. Royal Soc. Med. 9:295-300 (1965).

[17]By definition, an agent that creates a health risk of 2.0 is said to double the risk. Similarly, an agent that creates a health risk of 1.2 is said to increase a risk by 20 percent; and so on.

does not deem an agent or condition that is associated with a risk of less than 2.0 to be more likely than not the cause of an individual's injury, even though there is evidence that the agent or condition does increase the risk. Some courts use the same standard for admissibility of evidence. Consequently, in these courts unless an agent is associated with a relative risk greater that 2.0, the expert will not be permitted to testify about the possibility that the agent causes the illness or harm.

One speaker noted that the use of such "bright lines" is troubling to scientists for several reasons:

- There is no biological reason for the use of a 2.0 standard as opposed to any other standard.
- The use of such a hard standard obscures the fact that *any* risk means that some people could be harmed by the agent in question. For example, the relative risk of "passive" smoking—a person who never smoked but lives in proximity to smokers—is reliably shown to be about 1.2, i.e., the risk of developing lung cancer is elevated about 20 percent by passive smoking. A court using the standard of 2.0 would not admit evidence about passive smoking, and a group of people who have contracted lung cancer from passive smoking would not be able to seek compensation.
- In terms of verdicts, the plaintiff will lose every case and will collect nothing where the relative risk is 1.2, while the defendant will lose every case and will pay everyone where the relative risk is 2.2.
- The use of a bright line may confuse risk with incidence. Relative risk is the likelihood of the occurrence of a harmful effect, given certain assumptions about the frequency, duration, and magnitude of exposure. Public health authorities, however, usually consider the calculated incidence or frequency of an effect. For example, if a chemical is calculated to carry a risk at a rate of only one in 10,000, and only 10 people are exposed, there is no real expectation of an actual incidence of the effect.

Participants discussed several other points that may complicate court decisions. First, while the process of litigation may seek to assign risk to a single cause, most diseases are multi-factorial. Similarly, the "laws" of toxicology may be unclear to juries. For example, an overriding principle of toxicology is that the "dose makes the poison"; virtually everything can be toxic if taken in sufficient quantity. In addition, juries may not understand the concept of specificity. That is, each chemical may have a very specific effect that depends on its molecular structure; simply moving or exchanging a single atom may alter the effect of a substance. These scientific laws are easy to confuse, for example, when considering the toxicity of a substance, because the dosage of a toxic agent required to produce a particular disease in an animal may be far higher than the exposure of humans who are working with the toxic agent.

ASSESSING CAUSATION IN SPECIFIC INDIVIDUALS

The search for "truth" in the courtroom may become problematic when causative questions move beyond hard science to behavioral or philosophic realms. Physicians, in particular, are commonly asked to comment on issues that are "outside the box of scientific rigor," as one speaker said. At the bedside and in the clinic, diagnosis of signs and symptoms is done with a variety of therapeutic ends in mind. The first goal is to ameliorate suffering and disability. Intervention can include the treatment of the disease, the repair of injury, or the reduction or elimination of environmental causes of illness. A second goal is to comfort patients and their families. Here, the speaker stated, diagnosis is done largely for purposes of explanation—"as a narrative to hold on to." In the courtroom, the goal of diagnosis is to identify a cause of a condition for purposes of ascribing responsibility to people and institutions. Yet the legal system's desire to answer questions of responsibility and accountability frequently forces non-therapeutic questions on physicians. In child custody cases, for example, a physician may be asked which parent is "better" for the child, even when neither carries a disabling clinical diagnosis.

Criminal law moves even farther from clinical matters into questions of personal responsibility and moral or existential questions. Physicians (as well as psychiatrists and psychologists) may be asked to comment on the level of criminal intent, competency to stand trial, or the appropriateness of a death sentence, all of which are distant from the physician's familiar terrain of clinical diagnosis.

Remarkably, said one attorney, the scrutiny of medical experts specified by *Daubert* is rare. "When medical necessity is at stake in health insurance coverage cases," he said, "the plaintiffs' and the managed-care organizations' experts are almost always allowed in." Similarly, when the insanity defense is at issue in criminal cases, clinical experts are permitted to opine not merely about psychiatric diagnosis and symptoms but also about perceived responsibility of the examined individual—"without *Daubert* or *Kumho* standing in the way."

5

Can Scientists and Lawyers Get Along?

"One of the questions that we ought to be considering is whether . . . [the] current system of toxic tort litigation works given these very different objectives that we have, or whether one could somehow design a system that perhaps deals better with the uncertainties in both science and the law."

— Margaret A. Berger

It is not surprising to find that communication between legal and scientific experts in the courtroom is difficult. As a judge told the workshop, law and science are in some ways "about as different as they can be." He enumerated some differences: (1) law is backward-looking and its findings are based on precedent. If the judiciary had a bumper sticker, said the judge, it would say "Nothing new under the sun." Science is forward-looking in assuming that the truth is not yet fully known; (2) propositions to be tested in science are predictive; experiments are designed to be repeatable. Theories to be proved in court arise out of situations that occurred in the past and cannot be repeated; (3) scientists recognize that "truth" is mutable and may evolve over long periods of time. In adjudication, "truth," at least for the limited purpose of resolving disputes, must become final in a relatively short time; (4) scientists are often uncomfortable in the courtroom, especially during cross-examination. The dialogue is controlled by the lawyers and the judge, not by the expert witness; (5) disputes in science are resolved over time by peer review and the scrutiny of the scientific community at large. Legal disputes are resolved by cross-examination. Courts are not designed to determine whether scientific conclusions are correct.

Participants discussed further comparisons between peer review and cross-examination. Research performed for the purpose of litigation may afford no opportunity for true peer review, especially if the results were obtained recently. One lawyer noted that cross-examination can perform this function and can itself be an incentive to perform good, transparent science, and can even drive science forward. He provided the example that DNA identification litigation had driven DNA technology standards for laboratories.

LEGAL AND SCIENTIFIC VIEWPOINTS ON CAUSATION

In trying to determine whether there is a causative connection between Product A and Health Effect B, the *Daubert* trilogy directed the courts to look to the underlying foundation of the scientific testimony. Consequently, said an academic legal expert, the courts sometime seem to assume that the scientific community and the courts are examining the same concept of causation. When courts see that evidence is inconclusive from a scientific perspective they may decide that it thereby fails the Trilogy standards and should be excluded. This is a simplistic formulation, said the scholar, because it neglects to take into account some of the qualities of science and how they differ from those of the law.

Further, the speaker noted, it would be a mistake to assume that science and law are answering the same questions when asked to determine causation. The scientific process operates by testing hypotheses and rejecting those that are inconsistent with the data. The court system may exclude valuable knowledge from the deliberations when it excludes the results obtained in testing inconclusive hypotheses. The courts need to ask themselves, she said, whether such stringent scientific standards make sense in the legal setting.

The legal process approaches causation in a different way. A not-proven verdict in the courtroom has a clear significance. If one party fails to make a convincing case to the jury, it does not mean that more research should be done or that an assertion should be improved. It means that the party loses the case.

A "LIKELIER THAN NOT" STANDARD

On many issues of causation in tort law, pointed out a lawyer, there is scientific uncertainty. Yet in order to win the plaintiff does not have to prove beyond a reasonable doubt that the defendant caused the injury; the cause must only be "likelier than not."

One lawyer urged scientists to come up with a similar standard. "If there is one core thing that we lawyers would love to have from scientists," he said, "it is a better education of us and of judges on, if you will, a 'science of likelier than not.'" Such a science would clarify how much and what kinds of evidence would permit a scientific expert to make an educated guess about what is "likelier than not." He suggested that physicians must do this all the time because they have no choice. Physicians often do not know how to treat a patient, but they have to choose the best answer and treat accordingly. This is what the law would like from scientific testimony.[18]

Several participants discussed the idea of "relative allocation" or proportional recovery. This is done in comparative negligence cases, where, for example, multiple manufacturers have contributed to a chemical or waste spill whose effects have a long latency period. Similarly, suggested one judge, rather than requiring a bright-line, yes-or-no decision in every case based on scientific evidence, the law should be allowed to allocate responsibility and force a manufacturer to pay half of the cost, say, of a future insurance policy to clean up the spill. This would be an attempt to move the law in the direction of recognizing the uncertainties of risk and expert evidence.

An industry representative reminded the workshop participants not to look at pieces of evidence in isolation. He pointed out that the law no longer requires some of the elements of traditional tort in product liability cases. The balance has already shifted, he said, in favor of plaintiffs who no longer have to prove as many elements.

A BALANCE BETWEEN "TRUTH" AND USEFUL INFORMATION

The workshop returned to the different approaches of science and law as they try to determine causation. First, said a social scientist, courts err in the view that scientific studies invariably rest on some verifiable truth that is being determined in studies. At the same time, a study that is not conclusive in a scientific sense should not necessarily be ruled out of consideration by a court. For example, courts that reject opinions based on epidemiological studies that fail to satisfy a .05 level of statistical significance are losing potentially valuable information. What the "correct" level of significance is, is unknown. Various areas of science have adopted certain standards of proof to achieve the objectives and policies of those

---

[18]Developing such a standard will require collaboration between law and science. The science that is required, and therefore promoted, by the requirements of the law is only as good as the institutionally or statutorily imposed criteria for it.

disciplines. Those objectives may or may not be consistent with the objectives of the legal system and should be assessed against the policies of the legal forum.

Second, said a legal scholar, many courts assume that all the components of a scientist's opinion on causation are the products of empirically validated hypotheses. In fact, many such conclusions rest on conventions or models about which there is disagreement in the scientific community. This over-simplified view of science has increased the ability of federal trial judges to exclude parties' experts by ignoring the objectives of science and the law and by failing to examine the premises on which some scientific conclusions rest. And some courts have moved beyond science by creating new rules in the name of science that do not exist in the scientific community.

Speakers who believe that some of the emerging standards are too restrictive offered the following three examples:

- Epidemiology studies v. animal studies: Some courts have created a hierarchy of proof by insisting on epidemiological evidence. Insisting on epidemiological proof has policy implications, because epidemiological studies require more time—often years of exposure—and more money than animal studies. Because of the huge number of chemicals in use, it is rare that epidemiological studies of any single chemical are conducted. While it is true that evidence from animal studies must be used with great care, it is equally true that animal evidence should not be rejected out of hand without considering the context of the case.
- The rule of applying a relative risk standard of 2.0, as discussed earlier: This rule assumes that background risks are independent of the risks posed by the substance in question and can be calculated separately. Some scientists would reject such a model as inconsistent with a multi-factorial theory of disease.
- Some courts are imposing a rigid burden on plaintiffs to show the level of exposure; without showing this specific level, they cannot win. Two examples of cases in which the plaintiff will not be able to show precise levels of exposure are a condition that requires a long latency period to develop after initial exposure and a condition that results from continued exposure over a long period of time.

A law professor stated that the courts do handle causative uncertainty as a matter of course. In criminal law, he said, the court can decide that the defendant is "guilty beyond a reasonable doubt" and in tort law the court can make a "likelier-than-not" decision. A decision that would require a very high level of scientific certainty or probability, he said, exceeding likelier than not, would change the substantive law of torts. Similarly, requiring a 95 percent level of confidence that a probability estimate demonstrate an effect is an arbitrary custom chosen for the research community, but it is not necessarily dispositive in risk cases.

6

The Ethics of Expert Testimony

"... in thinking about an ethical framework, we may do well to transfer our attention from an individual-focused system that sees the problem as how to hold individuals to proper standards of behavior, to thinking instead [of] what kind of communal behavior is it for which we're trying to develop norms."
—Sheila Jasanoff

At first glance, it would seem that expert testimony should first of all adhere to a code of traditional ethics: thou shalt tell the truth, thou shalt not be intentionally inaccurate, and so on. It would also seem that the courts must enforce a standard of proper decorum and civil treatment of witnesses.

HOW TO TELL THE SCIENTIFIC TRUTH

Workshop participants struggled over the ethical questions of disclosure and scientific truthfulness. How does one define, enforce, and interpret full scientific disclosure? It is only possible to "tell the truth," after all, if there is general agreement about the definition of that truth. One participant said, half-jokingly, that "experts rarely tell the truth and certainly not the whole truth."

Beneath the humor lies a genuine difficulty, part of which stems from the courtroom atmosphere itself. In the pressure of the courtroom environment, said a scientist, experts are often tempted to give definitive answers where qualified ones would better fit the data. Another scientist said that experts may—perhaps unintentionally—neglect disconfirming data or the existence of other reasonable schools of thought. A legal scholar added that they also may exaggerate the significance of their own inferences and even forget, as a brief for *Daubert* reminds us, "that in science accepted 'truth' is not a constant: that it evolves, either gradually or discontinuously."[19]

One participant commented, "We all want in the legal system increased transparency about the way in which expert knowledge is produced, and a transparency that lets us get at right-wrong issues behind the practices and the testimony of experts." At the moment, he added, "we're in the early days of doing this kind of work and understanding the communal standards to which either the science side or the law side ought to subscribe."

One benefit of disclosure is that it creates a historical record about the provenance of research. With a "population" of studies about a given theme, one can see any association between the funding of the research and the outcome. One can also compare the outcomes of clinical trials that are funded by industry with trials that are funded by government.[20]

THE MYTH OF VALUE-FREE SCIENCE

Another central question concerns the objectivity of science itself. A lawyer reinforced the argument that the evidence of scientists is not "value-free": the testimony of a scientist may or may not be consistent with those of the legal system. When judges and attorneys uncritically accept the validity of an established scientific paradigm, he said, they make the same mistake as uncritical scientists.

AN ASYMMETRY OF DISCLOSURE REQUIREMENTS

A scientist raised the point of "asymmetry" in disclosure. That is, the expert witness has to disclose everything about which he or she is asked. Attorneys, on the other hand, can hire consultants, receive a report, and then decide whether and how to use the data collected in the report. Such data are said to be protected by the work product doctrine and therefore not discoverable by the opposing party. "This is natural to the adversary system," said a public-interest lawyer, "but I think it leads to less credibility."

---

[19] *Amici curiae*, Bloembergen et al., 18.

[20] The issues of sponsorship and design and control of research studies and outcomes is discussed in Chapter 7.

A CALL FOR GREATER TRANSPARENCY

Despite general agreement on the need for more transparency of expert information, including knowledge of all available information that bears on a question, most participants seemed reluctant to move toward formal standards for expert testimony so as not to unduly restrict the process of truth-seeking. On the other hand, a medical researcher said that the growth of specific standards for behavior in other well-specified situations had helped. He said that the field of medical research was better off for having set reasonable rules governing disclosure of conflict of interest, protection of human subjects, authorship, and research integrity, and that similar standards might be useful in the courtroom.

SOME FAILINGS OF EXPERT WITNESSES

A psychologist offered a critique of expert witnesses in his own field of psychology. Among the failings he found were:

- The use of psychological tests that had not been validated for the purpose at hand;
- Doctrinaire commitment to preconceived ideas;
- Forming initial impressions too quickly and failing to change these impressions in the face of new evidence.

"Court-appointed experts," he concluded, "as well as hired guns, may possess their own biases and foibles."[21]

A social scientist added to this comment by saying that the fault may be "much more systemic." She advocated a broader approach to ethics that takes full account of all the forces affecting the court system. "We can't limit ethical thinking to the behavior of particular expert witnesses in the courtroom," she said. "Although that's a place important to focus on, it should not be the only place."

---

[21]The opposing side can play a role in correcting these failures through competent cross-examination.

7

Scientific Research in the Context of the Legal System

"The cross-examination process is the tool that we utilize in the courtroom along with the advocacy of lawyers. What I think is missing from some scientific research that's done for purposes of litigation is the opportunity for there to be a true peer review process."

— Marsha Rabiteau

EYEWITNESS IDENTIFICATION RESEARCH

A social scientist described a notable—and rare—success story in which a psychological insight led to substantial change in the operation of the legal system—in this case, police lineup procedures. The insight grew out of eyewitness identification research that began in the 1970s. The problem was that the identification of guilty parties by eyewitnesses had great credibility in the courtroom even though the error rate of mistaken identification was high.

Psychologists attempted without success to convince the justice system of the unreliability of the technique through the 1980s and early 1990s. As long as 25 years ago, however, extensive psychological research with eyewitness identification had begun to expose the weaknesses of the technique. Psychologists found that witnesses tended to point to the person in the lineup who looked most like the perpetrator *relative to the other people in the lineup*—but who was not necessarily the perpetrator. They made a relative, not an absolute, judgment. Because someone in a lineup always looks more like the perpetrator than the other people in the lineup, witnesses tend to choose that person rather than choosing no one. By using a sequential system, in which witnesses were shown a single picture at a time without knowing whether they would see any more, the accuracy rate rose significantly.

Even after this finding had been established, however, the court system resisted any changes. Psychologists tried explaining it directly to the police, testifying as expert witnesses in court, and talking to the media. But general change did not begin until DNA testing arrived in the 1990s. It soon became apparent that DNA evidence had the capacity not only to convict but to exonerate those convicted mistakenly. Criminal justice researchers were able to show that 84 percent of these mistaken convictions were based on eyewitness evidence. In 1997, Attorney General Janet Reno, having seen this evidence, ordered the National Institute of Justice to develop the first set of national guidelines on eyewitness evidence and to include the substantial body of findings regarding eyewitness evidence produced by research psychologists. This set in motion the substantial change that has now taken place.

The social scientist listed several factors that finally led to change: (1) the scientists had clear experimental evidence; (2) they had publicized that evidence in leading, peer-reviewed psychology journals; and (3) they developed their own policy recommendations based on evidence, learned which policy makers could effect change, and lobbied those policy makers for change.

RESEARCH DESIGNED TO INFLUENCE COURT PROCEEDINGS

The session moderator, a scientist, suggested that courts may ask several questions about research whose results are used to influence court proceedings. First, does the content of the research meet the standards for scientific evidence discussed in *Daubert*? Second, is the provenance of the evidence appropriate: Is the organization or the person who did the research trustworthy? In what context was it issued?

The moderator went on to say that experts bringing evidence to trials may gather their information in two ways. The first is "before-the-fact" science, where the literature in the field is surveyed and presented by the expert. The second is "after-the-fact" science in which actual experiments or other research is done to answer a question in the context of particular litigation (such as the breast implant controversy). After-the-fact science has several virtues. It is relatively frugal, because the goal is clear and specific, and it is highly focused, so that the scientist can design the research to answer a specific question.

DOES SPONSORSHIP OF RESEARCH CAUSE BIAS?

Research that is sponsored to influence litigation, however, is regarded by the public and by segments of the research community with some cynicism. An example discussed at the workshop was the Women's Health Professionals Study, which produced large amounts of valuable information. The study drew a firestorm of attacks in the press when the *New England Journal of Medicine* revealed that one of the researchers had been paid as a consultant by four legal firms representing two manufacturers of breast implants. The press also revealed that Brigham Women's Hospital had received grants from Dow Corning to study silicone breast implants in a separate study. The publicity stimulated questions about whether sponsorship had compromised the integrity of the research. Workshop participants noted that even without evidence of bias, such controversies can pressure some research institutions to limit the funding contributions they will accept from industry, which often means that needed research is not carried out.

INDUSTRY-SPONSORED RESEARCH

A corporate lawyer said that industry has many incentives to maintain high standards of objectivity in the research it sponsors. "Industrial science has a certain transparency," she said. "Cooperation is key. Safety and good science are the friends of profit." The consequences for industry of either under- or over-reporting data can be extremely expensive, she pointed out. Therefore a code of "Good Laboratory Practices" has been in force since 1979 for many research-intensive industries that mandates a variety of measures, including retaining raw data about product development for 10 years or for the life of any resulting product. In addition, she said, enforcing federal agencies (such as the EPA or the FDA) tend to monitor industry studies to ensure that they are designed properly to answer questions about safety and other pertinent issues.

Another participant described how industry-sponsored drug tests for the Food and Drug Administration are performed. "Admitting new products is one of the most critical regulatory decisions we make as a nation," he said, "and it's critical to get our studies right on safety and efficacy." Drug studies are financed by the same prescription drug companies who intend to profit from the drugs; a similar process is followed for decisions about medical devices, food additives, and animal drugs. Because the format of these double-blind, placebo-controlled studies is standardized and because the FDA is often involved in the design of the studies, the agency has built up sufficient expertise and experience to produce objective and reliable results. For example, companies are required to examine all existing studies on the topic and submit the results of those studies whether they are favorable or not. (One participant pointed out that such a sweeping approach is virtually impossible to duplicate in a judicial setting.)

THE DIFFICULTY WITH TOXIC CHEMICALS

One participant discussed the difficulty in finding objective information about toxic chemicals. Few organizations or individuals can afford to do independent studies of human populations exposed to low levels of the thousands of chemicals that are potentially toxic, so there is little independent information in the literature. The need to know more about such chemicals usually arises in reaction to a lawsuit in which an expert, perhaps using animal studies or workplace studies, begins to formulate opinions. In rare cases, the plaintiff may fund a study, but the litigation context of the testimony may bar the plaintiff's expert evidence from admissibility.[22]

"The litigation context of the research is a factor that goes to weight," said the participant, "but I don't think it belongs on the admissibility scale. If it did, then the expert witness for the industry ought to be able to be excluded for the same reason." To balance the scale, he suggested, courts should require full disclosure of all possible biases, including the studies that were performed, that were not performed, and that were performed but rejected. Under present rules, the workshop participants were told, the opposing party has no discovery rights to reveal what parts of a particular study are being reported and what parts are not being reported.

STRUCTURING RESEARCH FOR LITIGATION

An engineer described his own initial reluctance to serve as an expert witness and said that he had decided to testify because of the need to "get credible technical testimony into the courtroom." He said that the cases had differed widely in subject matter and type of procedure. Even though he accepted funding from litigants for research on several specific questions, he felt that his work was impartial, his results interesting, and his testimony consequential as evidence. He did learn that scientific research

---

[22]Companies do perform some pre-production testing on many chemicals in an effort to avoid future product liability or federal regulation. Note the current "voluntary" programs to screen 1,000 High Production Volume Chemicals that might be endocrine disrupters or otherwise threaten the health of children.

standards are not always congruent with court admissibility standards, but he felt that the experience had been useful in providing important information for the courts, establishing objective technical testing on a controversial issue, and in learning more about the intersection of the legal and scientific worlds.

He added that he did not consider himself to be a "hired gun" or a professional witness, having participated in just three trials and one arbitration in 30 years. He suggested a series of simple principles by which experts could both participate meaningfully in litigation and also maintain their integrity:

- Design and perform the work on your own professional and ethical terms
- Publish all results in peer-reviewed journals, irrespective of the outcome
- Strive for equal objectivity whether working for the plaintiff or the defendant.

Another participant underlined the importance of doing one's best science for use in the courtroom. One reason is that cross-examination will tend to expose any weaknesses in methodology or conclusions. Another reason, of more general consequence, is that the effects of litigation may go beyond the finding of facts in specific cases to ultimately influence public policy. Although the criteria required for regulation are often more demanding than those required for admissibility of evidence, the public discussion of issues during litigation may prompt more extensive investigation by researchers.

8

Epilogue: Science and the Law— Toward Common Ground

"... science can sometimes get it wrong. ... Science is not always value-free, and most importantly, ... when lawyers accept the validity of an established scientific paradigm uncritically, they do risk making the same mistake as scientists."
—Michael Hoeflich

Several scientists and engineers who had served as expert witnesses concluded by urging their colleagues to enter the courtroom—both for their own education and to assist in this communication process. Several experts emphasized that the courtroom is one forum where scientists can help raise the public's understanding of science and the ways in which it underlies many aspects of modern life.

Appendixes

Appendix A:

Science, Technology, and Law Panel Biographies

Co-Chair: Donald Kennedy, (NAS/IOM), Ph.D. (Biology), Harvard, Bing Professor of Environmental Studies and Co-Director, Center for Environmental Science and Policy, Institute for International Studies, Stanford University. He is President Emeritus of Stanford University. He serves as Editor-in-Chief, *Science*. Previously, he served as Commissioner of the U.S. Food and Drug Administration. He was a member of the NAS planning committee that initiated the 1997 Academy Symposium on Science, Technology, and Law.

Co-Chair: Richard A. Merrill, (IOM), L.L.B., Columbia University School of Law, Daniel Caplin Professor of Law and Sullivan & Cromwell Research Professor of Law at the University of Virginia Law School. From 1975 to 1977 he served as Chief Counsel to the U.S. Food and Drug Administration. He was Dean of the University of Virginia Law School from 1980 to 1988. Since 1991 he has been special counsel to Covington & Burling. He was a member of the NAS planning committee that initiated the 1997 Academy Symposium on Science, Technology, and Law.

Frederick R. Anderson, J.D., Harvard Law School, heads the Energy, Environment, and Natural Resources Practice Group at Cadwalader, Wickersham & Taft in Washington, D.C. He is a former Dean of the Washington College of Law at American University. He was a member of the NAS planning committee that initiated the 1997 Academy Symposium on Science, Technology, and Law. Among his many publications is "Science

Advocacy and Scientific Due Process," *Issues in Science and Technology*, Summer 2000, pp. 71-76.

Margaret A. Berger, J.D., Columbia University, Suzanne J. and Norman Miles Professor of Law at Brooklyn Law School in Brooklyn New York. She has written extensively on science and law, and in particular on three key Supreme Court cases (Daubert, Joiner, Kumho) dealing with evidence. She is the co-author of Weinstein's *Evidence*.

Paul D. Carrington, L.L.B., Harvard, Harry R. Chadwick Senior Professor at Duke University Law School. He is the former Dean of Duke's Law School and has taught and published extensively on civil procedures. He was Reporter to the Advisory Committee on Civil Rules of the Judicial Conference of the United States. He also established the Private Adjudication Center that developed a Registry of Independent Scientists to provide disinterested advice to lawyers and judges on scientific issues that are the subject of legal disputes.

Joe S. Cecil, Ph.D. (Psychology) and J.D., Northwestern University, Project Director, Program on Scientific and Technical Evidence, Division of Research, Federal Judicial Center, Washington, D.C. He is responsible for judicial education and training in the area of scientific and technical evidence and the lead staff of the Federal Judicial Center's Scientific Evidence Manual, which is the primary source book on evidence for federal judges.

Joel E. Cohen, (NAS), Dr.P.H. (Population Sciences and Tropical Public Health) and Ph.D. (Applied Mathematics), Harvard, Abby Rockefeller Mauze Professor and Head, Laboratory of Populations, The Rockefeller University and Professor of Populations, Columbia University, in New York City. From 1991 to 1995, Dr. Cohen served as a U.S. Federal Court-appointed neutral expert on projections of asbestos-related claims associated with the Manville Personal Injury Settlement Trust. In addition, he has served as a Special Master in silicone gel breast implant products liability.

Rebecca S. Eisenberg, J.D., Professor of Law at the University of Michigan in Ann Arbor, Michigan. She teaches courses in intellectual property and torts and has taught on legal regulation of science and on legal issues associated with the Human Genome Project.

David L. Goodstein, Ph.D. (Physics), University of Washington, Vice Provost and Professor of Physics and Applied Physics, and Frank J. Gilloon

Distinguished Teaching and Service Professor at the California Institute of Technology. His book, *States of Matter*, helped launch a new discipline, condensed matter physics. In recent years, he has been particularly interested in societal issues that affect science as a profession.

Barbara S. Hulka, (IOM), M.D., Columbia College of Physicians and Surgeons, Kenan Professor, Department of Epidemiology, School of Public Health, University of North Carolina at Chapel Hill. Dr. Hulka's current research activities are in the field of cancer epidemiology—breast, uterine and prostate—and the application of biological markers. Dr. Hulka is working on the development of a process for incorporating scientific data into the judicial system.

Sheila Jasanoff, Ph.D., Harvard, J.D., Harvard, is Professor of Science and Public Policy at Harvard University's John F. Kennedy School of Government and the School of Public Health. Jasanoff's long-standing research interests center on the interactions of law, science, and politics in democratic societies. She is the author of numerous papers and books including, *The Fifth Branch: Science Advisors as Policymakers* and *Science at the Bar: Law, Science, and Technology in America*.

Robert E. Kahn, (NAE), Ph.D. (Electrical Engineering), Princeton University, is Chairman, CEO and President of the Corporation for National Research Initiatives (CNRI), a not-for-profit organization that provides funding and leadership to the research and development of the National Information Infrastructure. Dr. Kahn is a co-inventor of the TCP/IP protocols and a recipient of the 1997 National Medal of Technology awarded by President Clinton.

Daniel J. Kevles, Ph.D. (History), Princeton, is the Stanley Woodward Professor of History, at Yale University. Prior to this he was the Koepfli Professor of Humanities and directed the Program in Science, Ethics, and Public Policy at the California Institute of Technology in Pasadena, California. He has written extensively on issues regarding science and society including genetics, patenting, and scientific misconduct.

David Korn, (IOM), M.D., Harvard, Senior Vice President for Biomedical and Health Sciences Research, Association of American Medical Colleges, Washington, D.C. Previously, he served as Dean, Stanford University School of Medicine.

Eric S. Lander, (NAS/IOM), D.Phil. (Mathematics), Oxford University, Member, Whitehead Institute for Biomedical Research, Professor of Biology,

MIT, Director, Whitehead Institute/MIT Center for Genome Research, and Geneticist, Massachusetts General Hospital, Massachusetts Institute of Technology in Cambridge, Massachusetts. He is a geneticist, molecular biologist, and a mathematician, with research interests in human genetics, mouse genetics, population genetics, and computational and mathematical methods in biology. He has taught in the area of management and economics. Dr. Lander is a member of the American Academy of Forensic Sciences and has written about DNA Fingerprinting and other issues of science and law.

Patrick A. Malone, J.D., Yale Law School, partner, Stein, Mitchell & Mezines in Washington, D.C. Mr. Malone, a former medical journalist, represents plaintiffs in medical malpractice and product liability lawsuits. He is a member of the Association of Trial Lawyers of America and Trial Lawyers for Public Justice.

Richard A. Meserve, Ph.D. (Applied Physics), Stanford, J.D., Harvard, Chairman, Nuclear Regulatory Commission. Prior to his appointment he was a partner with the Washington, D.C., firm Covington and Burling, where he represented a number of corporate and non-profit organizations. He was a member of the NAS planning committee that initiated the 1997 Academy Symposium on Science, Technology, and Law. He wrote the amicus briefs on behalf of the National Academy of Engineering in the Kumho case and on behalf of the National Academy of Sciences in the Daubert case. These landmark cases established the basis for admitting expert testimony into court.

Alan B. Morrison, L.L.B., Harvard Law School, Director, Public Citizen Litigation Group, Washington, D.C. Public Citizen, Inc., is a non-profit citizen research, lobbying and litigation organization founded in 1971 by Ralph Nader.

Harry J. Pearce, J.D., Northwestern University School of Law, is Chairman of Hughes Electronics Corporation, a subsidiary of General Motors Corporation in El Segundo, California. He previously served General Motors as Vice Chairman, and prior to that as General Counsel. Mr. Pearce has been admitted to the U.S. Supreme Court, U.S. Court of Military Appeals, Eight Circuit Court of Appeals, various U.S. District Courts and State District Courts, and the Michigan Supreme Court.

Henry Petroski, (NAE), Ph.D., University of Illinois, A.S. Vesic Professor of Civil Engineering and Professor of History, Duke University, in Durham, North Carolina. He has written extensively on the nature of engineering and design, as well as on engineering and law issues. Most recently, he authored a chapter on engineering practice for the Federal Judicial Center's evidence project.

Channing R. Robertson, Ph.D. (Chemical Engineering), Ruth G. and William K. Bowes Professor, School of Engineering, and Professor, Department of Chemical Engineering, Stanford University. Dr. Robertson conducted research on several products about which there was extensive litigation and in which he served as an expert witness.

Pamela Ann Rymer, L.L.B., Stanford, Circuit Judge, U.S. Court of Appeals for the Ninth Circuit, Pasadena, California. She was appointed in 1989 by President George Bush. Judge Rymer currently serves as the Chair of the AAAS Court-Appointed Scientific Experts Demonstration Project.

STAFF OF THE SCIENCE, TECHNOLOGY, AND LAW PROGRAM

Anne-Marie Mazza, Director. Dr. Mazza joined the National Academies in 1995. She served as Senior Program Officer with both the Committee on Science, Engineering, and Public Policy and the Government-University-Industry Research Roundtable. From 1999 to 2000, she divided her time between the STL Program and the White House Office of Science and Technology Policy, where she chaired an interagency working group on the government-university research partnership. She received a Ph.D. in Public Policy from the George Washington University.

Susie Bachtel, Staff Associate. Ms. Bachtel became a Staff Associate of the National Academies in 1998. She previously was Special Assistant to the Director, the White House Office of Science and Technology Policy, and prior to that Executive Assistant to the Director of the U.S. Office of Technology Assessment. Ms. Bachtel received a B.A. in Social Sciences from the Ohio State University.

Maarika Liivak served in the Science, Technology, and Law Program as a Christine Mizrayan Intern. After finishing her B.A. in biochemistry from Rutgers University, she undertook her graduate studies with the support of a National Science Foundation Pre-Doctoral Fellowship at Yale University where she earned a M.S. in genetics. She will begin law school at Yale in the fall. She eventually would like to combine her training in genetics with a legal career.

Kevin Whittaker served in the Science, Technology, and Law Program as a Christine Mizrayan Intern.

Kirsten A. Moffatt, Ph.D., Consultant. Dr. Moffatt received her Ph.D. in Experimental Pathology from the University of Colorado Health Science Center. Her thesis work focused on the molecular mechanism(s) through which vitamin D acts to decrease the growth of prostate cancer cells.

Alan H. Anderson is a consultant writer who has written National Academy reports on a variety of topics, including science policy, graduate and postdoctoral education, capitalizing on the results of research, the cyclicality of the semiconductor industry, the "new economy," women in science and engineering, and the Government Performance and Results Act. He also writes for the Institute for Advanced Study, in Princeton, N.J., and other clients. He has been a science writer for *Time* magazine and other publications, and holds a master's degree from the Columbia University School of Journalism and a B.A. in English from Yale University.

Appendix B:

Agenda

SCIENTIFIC EVIDENCE WORKSHOP

Science, Technology, and Law Program
The National Academies
2101 Constitution Avenue, NW
Washington, D.C.
September 7, 2000

8:00   *Continental Breakfast*

8:30   *Welcome and*
       Introductions: Richard A. Merrill, Daniel Caplin Professor of Law, University of Virginia Law School, and Co-Chair, Science, Technology, and Law Program

8:40   *How Does Scientific Expert Testimony Compare with Scientific Practice?*

       Moderator: Richard A. Levie, Principal, ADR Associates, L.L.C., Senior Judge, Superior Court of the District of Columbia (ret.)

45

Panelists:

Epidemiology: David Ozonoff, Professor and Chair,
Department of Environmental Health, School of
Public Health, Boston University

Jon Samet, Professor and Chair, Department of
Epidemiology, School of Public Health, The
Johns Hopkins University

Toxicology: Bernard D. Goldstein, Director, Environmental
and Occupational Health Sciences Institute,
University of Medicine and Dentistry of New
Jersey—The Robert Wood Johnson Medical
School

Differential
Diagnosis: M. Gregg Bloche, Professor of Law,
Georgetown University Law School and
Adjunct Professor of Public Health, Johns
Hopkins University

10:15   *Break*

10:30   *Emerging Issues*

Moderator: Joel E. Cohen, Abby Rockefeller Mauze
Professor, and Head, Laboratory of
Populations, The Rockefeller University and
Columbia University

*Issue One: To What Extent Are Evidentiary Rulings Becoming
Substantive Standards of Law?*

Presenter: Margaret A. Berger, Suzanne J. and Norman
Miles Professor of Law, Brooklyn School of Law

Commentators: Leon Gordis, Professor of Epidemiology, School
of Public Health, Professor of Pediatrics, School
of Medicine, The Johns Hopkins University

Michael H. Gottesman, Professor of Law,
Georgetown University Law Center

Issue Two: The Ethics of Expert Evidence

Presenter: Sheila Jasanoff, Professor of Science and Public Policy, John F. Kennedy School of Government, Harvard University

Commentator: Donald N. Bersoff, Professor of Law and Director of the Law and Psychology Program, Villanova School of Law

Issue Three: Using Social Science to Affect Policy: Eyewitness Research as a Successful Example

Presenter: Gary L. Wells, Professor of Psychology, Iowa State University

Commentator: Shari Diamond, Professor of Law and Psychology, Northwestern University Law School

12:45   *Lunch*   Gina Kolata, Science Desk, *The New York Times*

2:00   *Research Sponsored by Parties to Challenge, Support, or Influence Judicial or Regulatory Proceedings. How Does Sponsorship Affect Research and the Way It Is Perceived?*

Moderator: Donald Kennedy, Bing Professor of Environmental Sciences and President Emeritus, Stanford University, and Editor-in-Chief, *Science*, Co-Chair, Science, Technology, and Law Program

Panelists: Marsha Rabiteau, Counsel, The Dow Chemical Company

               Channing R. Robertson, Ruth G. and William K. Bowes Professor, Department of Chemical Engineering, Stanford University

               William B. Schultz, Deputy Assistant Attorney General, Civil Division, U.S. Department of Justice

4:00   **Break**

4:15   **Different Perspectives in Search of Mutual Understanding**

        Moderator: Paul D. Carrington, Harry R. Chadwick Senior Professor of Law, Duke University Law School

        Presentator: Michael H. Hoeflich, Dean and John H. and John M. Kane Professor of Law, University of Kansas Law School

        Commentators: David A. Freedman, Professor of Statistics, University of California, Berkeley

                          Patrick A. Malone, Stein, Mitchell & Mezines

5:15   **Closing
        Remarks**   The Honorable Andre M. Davis, Judge, U.S. District Court for the District of Maryland

5:30   **Reception**

Appendix C:

Scientific Evidence Workshop
List of Registrants

Ms. Jacqueline A. Ahn
Research Assistant
Institute for Law and Justice
Alexandria, VA

Professor Ronald J. Allen
Northwestern University School of Law
Chicago, IL

Mr. Frederick R. Anderson, Jr.
Cadwalader, Wickersham & Taft
Washington, DC

Dr. Martin A. Apple
President
Council of Scientific Society Presidents
Washington, DC

Ms. Kathryn S. Arnow
Health Policy Analyst
Washington, DC

Ms. Susie Bachtel
Staff Associate
Science, Technology, and Law Program
The National Academies
Washington, DC

Mr. William G. Barnes
Environmental Claims Director
St. Paul Fire & Marine Insurance Company
St. Paul, MN

Dr. David Z. Beckler
Carnegie Commission on Science, Technology and Government
Fairfax, VA

Professor Margaret A. Berger
Suzanne J. and Norman Miles Professor of Law
Brooklyn Law School
Brooklyn, NY

Professor Donald N. Bersoff
Professor
Law & Psychology Program
Villanova School of Law
Villanova, PA

Ms. Jacqueline Besteman
Senior Advisor
Center for Practice and Technology Assessment
Agency for Healthcare Research and Quality
U.S. Department of Health and Human Services
Washington, DC

Mr. Blake A. Biles
Senior Partner
Arnold & Porter
Washington, DC

Dr. Richard E. Bissell
Executive Director
Policy Division
The National Academies
Washington, DC

Professor M. Gregg Bloche
Professor of Law
Georgetown University Law Center
Washington, DC

Ms. Anne W. Bloom
Attorney
Trial Lawyers for Public Justice
Washington, DC

The Honorable Danny J. Boggs
U.S. Circuit Judge
Sixth Circuit Court of Appeals
Louisville, KY

Professor Eugene Borgida
Professor of Psychology and Law
Department of Psychology
University of Minnesota
Minneapolis, MN

Mr. Darren A. Bowie
Assistant Director
Federal Trade Commission
Washington, DC

Lynn A. Bristol, Ph.D., J.D.
Arent Fox Kinter Plotkin & Kahn, PLLC
Washington, DC

Mr. Duncan M. Brown
Consultant
Science, Technology, and Law Program
The National Academies
Washington, DC

Ms. Wanda V. Brownlee
William J. Brownlee, MDPC
Washington, DC

William J. Brownlee, M.D.
William J. Brownlee, MDPC
Washington, DC

Professor Paul D. Carrington
Harry R. Chadwick Senior Professor
Duke University Law School
Durham, NC

Ms. Russellyn S. Carruth
Attorney at Law
Piscataway, NJ

Mr. John A.C. Cartner
Principal
MSCL, Inc.
Alexandria, VA

Joe Cecil, Ph.D., J.D.
Project Director
Program on Scientific and Technical Evidence
The Federal Judicial Center
Washington, DC

Mr. Robert P. Charrow
Of Counsel
Crowell & Morning, LLP
Washington, DC

Mr. McAlister Clabaugh
Program Associate
Science, Technology, and Economic Policy Program
The National Academies
Washington, DC

Mr. William T. Clemons
Attorney
Engstrom, Lipscomb & Lack
Southern Pines, NC

Professor Joel E. Cohen
Abby Rockefeller Mauze Professor and
Head, Laboratory of Populations
Columbia University and
The Rockefeller University
New York, NY

Dr. E. William Colglazier
Executive Director
National Research Council
The National Academies
Washington, DC

Ms. Lillie B. Coney
Senior Special Assistant
Office of The Honorable Sheila Jackson-Lee
U.S. House of Representatives
Washington, DC

Professor Carl E. Cranor
Professor of Philosophy
University of California, Riverside
Riverside, CA

Ms. Maggie E. Daley
Consultant
CMBE
Washington, DC

Mr. B. Michael Dann
Visiting Fellow
National Center for State Courts
Williamsburg, VA

The Honorable Andre M. Davis
Judge
U.S. District Court for the District of Maryland
Baltimore, MD

Mr. Silvio J. DeCarlie
Corporate Counsel
DuPont Company
Wilmington, DE

Professor Shari S. Diamond
Professor
School of Law
Northwestern University
Chicago, IL

Dr. Sidney Draggan
Senior Science and Science Policy Adviser
U.S. Environmental Protection Agency
Washington, DC

Dr. Meghan A. Dunn
Research Associate
The Federal Judicial Center
Washington, DC

Professor Rebecca S. Eisenberg
Professor of Law
University of Michigan Law School
Ann Arbor, MI

Professor Lucinda M. Finley
Professor of Law
SUNY Law School at Buffalo
Buffalo, NY

Mr. Kevin Finneran
Editor-in-Chief
Issues in Science and Technology
The National Academies
Washington, DC

Dr. Eric A. Fischer
Senior Specialist in Science and Technology
Congressional Research Service
The Library of Congress
Washington, DC

Ms. Lisa Forman
Acting Director
Investigative and Forensic Sciences Division
National Institute of Justice
Washington, D.C

Dr. Mark S. Frankel
Director
Program on Scientific Freedom, Responsibility, and Law
American Association for the Advancement of Science
Washington, DC

Professor David Freedman
Professor of Statistics
University of California, Berkeley
Berkeley, CA

Mr. Jeffrey Freedman
Attorney
Office of the General Counsel
Nycomed Amersham Imaging
Princeton, NJ

Ms. Sharon T. Friedman
Senior Policy Analyst
Office of Science and Technology Policy
Executive Office of the President
Washington, DC

Professor William P. Gardner
Center for Bioethics and Health Law
Pittsburg, PA

Ms. Julie Spence Gefke
Senior Forensic Analyst
National Institute of Justice
Washington, DC

Ms. Jody George
Senior Judicial Education Specialist
The Federal Judicial Center
Washington, DC

Ms. Kristina H. Gill
Legal Assistant
Research Division
The Federal Judicial Center
Washington, DC

Karen A. Goldman, Ph.D., J.D.
Extramural Technology Transfer Policy Specialist
Office of Science Policy
Director's Office
National Institutes of Health
Rockville, MD

Dr. Bernard G. Goldstein
Director
Environmental and Occupational Health Sciences Institute
UMDNJ—Robert Wood Johnson Medical School
Piscataway, NJ

Ms. Susan Goodman
Director of Stewardship
Development Office
The National Academies
Washington, DC

Professor Leon Gordis
Department of Epidemiology
The Johns Hopkins University
Baltimore, MD

Professor Michael Gottesman
Professor of Law
Georgetown University Law Center
Washington, DC

Ms. Karen Gottlieb
Court Consultant
Bio-Law
Nederland, CO

Professor Jack Halpern
Louis Block Distinguished Professor
Department of Chemistry
University of Chicago
Chicago, IL

Ms. Tamar P. Halpern
Partner
Phillips Lytle Hitchcock Blaine & Huber
Buffalo, NY

Professor Craig Haney
Professor and Chair
Department of Psychology
University of California, Santa Cruz
Santa Cruz, CA

Mr. Henry R. Hertzfeld
Senior Research Scientist
Center for International Science and Technology Policy
The George Washington University
Washington, DC

Professor Michael H. Hoeflich
John H. and John M. Kane Distinguished Professor of Law
University of Kansas School of Law
Lawrence, KS

Ms. Alice S. Hrdy
Attorney
Federal Trade Commission
Washington, DC

Professor Barbara S. Hulka
Kenan Professor
Department of Epidemiology
School of Public Health
University of North Carolina, Chapel Hill
Chapel Hill, NC

Ms. Ann C. Hurley
Trial Attorney
Environment and Natural Resources Division
U.S. Department of Justice
Washington, DC

Professor Sheila Jasanoff
Professor of Science and Public Policy
John F. Kennedy School of Government
Harvard University
Cambridge, MA

Mr. Richard A. Johnson
Senior Partner
Arnold & Porter
Washington, DC

Ms. Rhonda M. Jones
Legislative Fellow
Office of the Honorable Sherwood Boehlert
U.S. House of Representatives
Washington, DC

Mr. Alan R. Kabat
Attorney
Bernabei & Katz, PLLC
Washington, DC

Dr. Bruce J. Kelman
Toxicologist/Managing Principal
GlobalTox, Inc.
Redmond, WA

Professor Donald Kennedy
Bing Professor of Environmental Studies
Co-Director
Center for Environmental Science and Policy
Institute for Environmental Studies
Stanford University
Stanford, CA

Professor Howard M. Kipen
Director and Professor
Environmental and Occupational Health Sciences Institute
UMDNJ—Robert Wood Johnson Medical School
Piscataway, NJ

Dr. Rebecca J. Klemm
President
Klemm Analysis Group, Inc.
Washington, DC

Ms. Gina Kolata
Writer
Science Desk
The New York Times
New York, NY

David Korn, M.D.
Senior Vice President for Biomedical and Health Sciences Research
Association of American Medical Colleges
Washington, DC

Dr. Sylvia K. Kraemer
Visiting Professor
School of Public Policy
George Mason University
Arlington, VA

Ms. Carol L. Krafka
Senior Research Associate
The Federal Judicial Center
Washington, DC

Professor Eric S. Lander
Professor of Biology and Director
Whitehead Institute
MIT Center for Genome Research
Massachusetts Institute of Technology
Cambridge, MA

Professor Richard O. Lempert
Professor of Law and Sociology
University of Michigan Law School
Ann Arbor, MI

The Honorable Richard A. Levie (Ret.)
Principal
ADR Associates, L.L.C.
Washington, DC

Mr. Scott P. Lockledge
Science Fellow
Office of The Honorable Vernon Ehlers
U.S. House of Representatives
Washington, DC

Mr. Emile Loza
Law Clerk
Internet Fraud Act Committee
Federal Trade Commission
Washington, DC

Mr. Malcolm N. MacKinnon, III
President
MSCL, Inc.
Alexandria, VA

Mr. Patrick A. Malone
Partner
Stein, Mitchell & Mezines
Washington, DC

Ms. Jennifer E. Marsh
Research Associate
The Federal Judical Center
Washington, DC

Ms. Rose Marshall
President
Legal PR, Ltd.
Arlington, VA

Ms. Lorelie S. Masters
Director
Beveridge & Diamond, PC
Washington, DC

Dr. Anne-Marie Mazza
Director
Science, Technology, and Law Program
The National Academies
Washington, DC

Mr. Robert A. McCarter
Attorney
Arnold & Porter
Washington, DC

Mr. Merrill Meadow
Senior Development Officer
The National Academies
Washington, DC

Professor Richard A. Merrill
Daniel Caplin Professor of Law
University of Virginia Law School
Charlottesville, VA

Dr. Steve Merrill
Director
Science, Technology, and Economic Policy Program
The National Academies
Washington, DC

The Honorable Paul R. Michel
Circuit Judge
U.S. Court of Appeals, Federal Circuit
Washington, DC

Dr. Wilhelmine Miller
Senior Program Officer
Institute of Medicine
The National Academies
Washington, DC

Mr. Ned I. Miltenberg
Senior Counsel
Association of Trial Lawyers of America
Washington, DC

Mr. Alan B. Morrison
Director
Public Citizen Litigation Group
Washington, DC

The Honorable Pauline Newman
Circuit Judge
U.S. Court of Appeals, Federal Circuit
Washington, DC

Professor Karen Nordeheden
Professor
Chemical and Petroleum Engineering Department
University of Kansas
Lawrence, KS

Professor David Ozonoff
Professor and Chair
Department of Environmental Health
School of Public Health
Boston University
Boston, MA

Kathleen A. Pajer, M.D., M.P.H.
Associate Professor of Psychiatry
Western Psychiatric Institute and Clinic
University of Pittsburgh
Pittsburgh, PA

Barry M. Parsons, Esq.
Associate
Cromwell & Moring, LLP
Washington, DC

Dr. Aristides Patrinos
Associate Director of Science for Biological and Environmental Research
Office of Biological Research
U.S. Department of Energy
Germantown, MD

Ms. Ellen Paul
Public Policy Representative
American Institute for Biological Sciences
Washington, DC

Ms. Judyth W. Pendell
Pendell Consulting
Bloomfield, CT

Dr. Carol V. Petrie
Director
Committee on Law and Justice
Commission on Behavioral and Social Sciences and Education
The National Academies
Washington, DC

Dr. Henry Petroski
Professor
Department of Civil & Environmental Engineering
Duke University
Durham, NC

Professor Susan Poulter
College of Law
University of Utah
Salt Lake City, UT

Dr. Donald Prosnitz
Chief Science and Technology Advisor
U.S. Department of Justice
Washington, DC

Professor Federick A. Provordy
Harold R. Tyler Professor of Law, and
Director, Science and Technology Center
Albany Law School
Albany, NY

Ms. Marsha Rabiteau
Counsel
The Dow Chemical Company
Midland, MI

Mr. Tim Reagan
Senior Research Associate
The Federal Judicial Center
Washington, DC

Ms. Holly E. Reed
Research Associate
Committee on Population
The National Academies
Washington, DC

Professor Channing R. Robertson
Ruth G. and William K. Bowes Professor
Department of Engineering
Stanford University
Stanford, CA

Dr. Joseph V. Rodricks
ENVIRON International Corp.
Arlington, VA

Professor Anthony Z. Roisman
Attorney
Hershenson, Carter, Scott & McGee
Norwich, CT

Mr. Ernie Rosenberg
President
Soap and Detergent Association
Washington, DC

Mr. David B. Rottman
Associate Director of Research
National Center for State Courts
Williamsburg, VA

Ms. Deborah Runkle
Senior Program Associate
American Association for the Advancement of Science
Washington, DC

Professor Jonathan Samet
Professor and Chair
Department of Epidemiology
The Johns Hopkins Univesity
Baltimore, MD

Mr. Tom Scarlett
Research Counsel
Association of Trial Lawyers of America
Washington, DC

Mr. Nathan A. Schachtman
Partner
McCarter & English, LLP
Philadelphia, PA

Mr. Craig M. Schultz
Research Associate
Science, Technology, and Economic Policy Program
The National Academies
Washington, DC

Mr. William B. Schultz
Deputy Assistant Attorney
General Civil Division
U.S. Department of Justice
Washington, DC

Mr. Benjamin Seggerson
Assistant Account Executive
Legal PR, Ltd.
Arlington, VA

Professor George F. Sensabaugh
Professor
School of Public Health
University of California, Berkeley
Berkeley, CA

Dr. Joseph F. Shelley
President
Techanalysis, Inc.
Princeton, NJ

Mr. Ethan Shenkman
Counsel to the Assistant Attorney Gnereal
Environment and Natural Resources Division
U.S. Department of Justice
Washington, DC

Mr. Ronald Simon
Attorney
Simon & Associates
Washington, DC

Mr. Frederick T. Smith
Partner
McCarter & English, LLP
Newark, NJ

Mr. Gerson Smoger
Principal
Smoger & Associates, P.C.
Oakland, CA

Kathryn E. Stein, Ph.D.
Director
Division of Monocolonal Antibodies
Center for Biologics Evaluation and Research
U.S. Food and Drug Administration
Bethesda, MD

Ms. Christina A. Studebaker
Research Associate
The Federal Judical Center
Washington, DC

Ms. Melissa L. Sturges
Program Associate
Court Appointed Scientific Experts Project
American Association for the Advancement of Science
Washington, DC

Ms. Anjali R. Swienton
Senior Forensic Analyst
Investigative and Forensic Science Division
National Institute of Justice
Washington, DC

Daniel T. Teitelbaum, M.D., P.C.
Medical Toxicologist
Denver, CO

Dr. Sheldon L. Trubatch, Esq.
Attorney
Foley & Lardner
Washington, DC

Dr. Myron F. Uman
Associate Executive Officer
National Research Council
The National Academies
Washington, DC

Mr. John Vail
Senior Counsel
Association of Trial Lawyers of America
Washington, DC

Professor Gary Wells
Professor
Psychology Department
Iowa State University
Ames, IA

Dr. Charles W. Wessner
Deputy Director
Science, Technology, and Economic Policy Program
The National Academies
Washington, DC

Mr. Kevin Whittaker
Syracuse University Law School
Syracuse, NY

Ms. Robin S. Wilson
Executive Assistant
National Commission on the Future of DNA Evidence
National Institute of Justice
Washington, DC

Mr. Richard L. Withers
Director
The Medical College of Wisconsin
Milwaukee, WI

Ms. Martha K. Wivell
Attorney
Robins Kaplan Miller & Ciresi
Minneapolis, MN

Mr. Brian Wolfman
Staff Lawyer
Public Citizen Litigation Group
Washington, DC

Dr. Gillian R. Woollett
Associate Vice President
Biologics and Biotechnology
Pharmaceutical Research and Manufacturers of America
Washington, DC

Mr. Miles J. Zaremski
Partner
Kovitz Shifrin & Waltzman
Buffalo Grove, IL

Mr. David F. Zoll
Vice President and General Counsel
American Chemistry Counsel
Arlington, VA